# Contents

GW00640795

# Introduction

This booklet is intended to help doctors and medical students, nurses and clinical chemistry laboratory workers understand SI units, and to provide conversion scales for SI units and conventional units for the tests most commonly used in clinical chemistry. Ranges are also provided to indicate values found in 95% of healthy adults, and of a hospital population. For some tests, we have included short comments about specimen collection or interpretation of results. These have been limited to points which in our experience cause most confusion.

# SI units

The International System of Units (Système International, SI) was adopted in 1960 by the General Conference of Weights and Measures as a logical, coherent system based on seven fundamental units: metre, kilogram, second, ampere, kelvin, candela and mole.

The system has been adopted generally by international scientific bodies, including the International Federation of Clinical Chemistry (I.F.C.C.) and the Section of Clinical Chemistry of the International Union of Pure and Applied Chemistry (I.U.P.A.C.).

All measurements are expressed in the basic units or in units derived from them. The seven basic units, and some of the derived units relevant to medicine with standard abbreviations are:

| Physical quantity | Name of SI unit | Symbol |
|---|---|---|
| length | metre | m |
| mass | kilogram | kg |
| time | second | s |
| electric current | ampere | A |
| temperature | kelvin | K |
| luminous intensity | candela | cd |
| amount of substance | mole | mol |
| energy | joule | J |
| force | newton | N |
| power | watt | W |
| pressure | pascal | Pa |

Prefixes to indicate fractions or multiples of the basic or derived units have also been defined.

| Fraction | Prefix | Symbol |
|---|---|---|
| $10^{-1}$ | deci | d |
| $10^{-2}$ | centi | c |
| $10^{-3}$ | milli | m |
| $10^{-6}$ | micro | $\mu$ |
| $10^{-9}$ | nano | n |
| $10^{-12}$ | pico | p |
| $10^{-15}$ | femto | f |
| $10^{-18}$ | atto | a |

| Multiple | Prefix | Symbol |
|----------|--------|--------|
| 10 | deca | da |
| $10^2$ | hecto | h |
| $10^3$ | kilo | k |
| $10^6$ | mega | M |
| $10^9$ | giga | G |
| $10^{12}$ | tera | T |

The major change in clinical chemistry is the reporting of analyses in mmol/l (or $\mu$mol/l, nmol/l, etc.) in place of the conventional mg/100 ml. For example, determinations reported in mmol/l may be converted to mg/100 ml thus:

$$\frac{\text{Conc. in mmol/l} \times \text{molecular weight of substance}}{10} = \text{mg/100 ml}$$

When the molecular weight of a substance cannot be accurately determined (as in mixtures), values are reported as weight of substance/litre, e.g. globulin will be reported as g/l. In all reports, the litre (l) is the preferred volume.

Another change is in reporting of pressure, mmHg being replaced by the derived unit the pascal ($Nm^{-2}$) or in practice since this is too small, the kilopascal (kPa).

The reporting of enzyme activities in biological fluids is not affected by the adoption of SI units. The recommended convention is I.U./l (1 I.U./l represents 1 $\mu$mol of substrate changed/minute).

It is important that doctors, nurses and laboratory workers become used to thinking in SI units as rapidly as possible. The conversion scales are provided to help staff to adjust to these new units, and have been deliberately designed to make the new easier than the old! While there is complete agreement on the precise conventions to be used for reporting most tests, there is still uncertainty about some tests, notably drug levels and pH.

**References**

Ellis G. (ed) (1971) *Units, Symbols and Abbreviations*. Royal Society of Medicine, London.

Baron D. N. *et al.* (1974) The use of SI units in reporting results obtained in hospital laboratories. *J. clin. Path.* **27**, 590–597.

# Normal ranges (reference values)

On each page 'normal' or reference values are indicated. The term 'normal' has specific statistical connotations, as well as less defined implications of health, that are not always appropriate. We prefer to use the term reference values. In general, the limits are those that enclose 95% of individuals.

**You should check with your local laboratory, if it uses methods producing similar values.** For many tests, similar values are given by any reputable analytical method. Differences between methods may be significant for tests such as albumin, calcium and glucose and attention is drawn to this in the notes. **Large** differences may occur in measured **enzyme** activity, since differences in substrate, temperature and other conditions affect the results **even** when reported in international units per litre (I.U./l).

The old concept of a 'normal' range based on analysis of samples from 100 or so blood donors or laboratory workers is outmoded for two reasons. First, an overall normal range obscures what may be significant changes with ageing or between sexes. Secondly, the population on which the normal range is based may be different from that being studied. One of the most important factors is posture; for example, serum proteins and protein-bound substances are up to 10 to 20% lower in a patient in bed than in the same patient when ambulant. The stress of hospitalisation may also cause changes in serum constituents, all too little documented. Other factors such as race, diet and social class further complicate the problem. It is however scarcely feasible to provide data from reference populations to cover all variables.

Much of the data on reference values presented in this booklet is derived from two main sources, from apparently healthy ambulant individuals taking part in screening programmes, and on profile analysis of several thousand inpatients at the Queen Elizabeth Hospital, Birmingham. Specimens were analysed at the BUPA Laboratories or the Clinical Chemistry Department at the Queen Elizabeth Hospital. For other tests we have only limited data of our own, and this has been combined with published data from other sources. It is unfortunate that information on reference values for some less common but important tests is limited and sometimes out of date.

Reference values on a hospital population are controversial. Since healthy people are not found in bed in hospital, it is not easy to distinguish the effects of disease, sometimes undiagnosed, on clinico-chemical values, from non-specific factors such as bed rest and stress. The values on hospital patients presented here were obtained after excluding patients with known renal disease, and all values in the lowest and highest 1%.

We believe that values on healthy outpatients are the best general guide, but that clinicians should be aware of what happens in a hospital population. Values outside the 'healthy' ambulant values but within those found in hospital inpatients should be treated with caution. Whether they should be ignored or further investigated will depend on the overall clinical problem. We do firmly believe that much more attention should be paid to changes with age and sex, and for many determinations adequate data are still not available.

How much attention should be paid to small differences from the appropriate reference range is a matter of clinical judgement. We have simply attempted to provide basic facts.

Throughout we have used serum as equivalent to plasma, except where attention is drawn to differences.

### References

The literature on normal ranges is enormous. We think the following brief list is relevant:

Amador E. (1973) Normal ranges. In Stefanini M. *Progress in Clinical Pathology*, Vol. 5, 59–83. Grune & Stratton, New York.

Wilding P., Rollason J. G. & Robinson D. (1972) Patterns of change for various biochemical constituents in well population screening. *Clinica Chimica Acta,* **41,** 375–387.

The data on hospital inpatients at the Queen Elizabeth Hospital are to be published by Professor T. P. Whitehead and his colleagues. We are grateful to Professor Whitehead for permission to include these data here.

# Serum albumin (g/l)

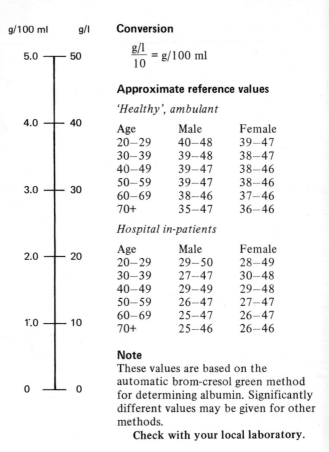

| g/100 ml | g/l |
|----------|-----|
| 5.0 | 50 |
| 4.0 | 40 |
| 3.0 | 30 |
| 2.0 | 20 |
| 1.0 | 10 |
| 0 | 0 |

**Conversion**

$$\frac{g/l}{10} = g/100\ ml$$

**Approximate reference values**

*'Healthy', ambulant*

| Age | Male | Female |
|-------|-------|-------|
| 20–29 | 40–48 | 39–47 |
| 30–39 | 39–48 | 38–47 |
| 40–49 | 39–47 | 38–46 |
| 50–59 | 39–47 | 38–46 |
| 60–69 | 38–46 | 37–46 |
| 70+ | 35–47 | 36–46 |

*Hospital in-patients*

| Age | Male | Female |
|-------|-------|-------|
| 20–29 | 29–50 | 28–49 |
| 30–39 | 27–47 | 30–48 |
| 40–49 | 29–49 | 29–48 |
| 50–59 | 26–47 | 27–47 |
| 60–69 | 25–47 | 26–47 |
| 70+ | 25–46 | 26–46 |

**Note**

These values are based on the automatic brom-cresol green method for determining albumin. Significantly different values may be given for other methods.

**Check with your local laboratory.**

# Serum bicarbonate (mmol/l)

**Conversion**
Nil. For bicarbonate, mmol/l = mEq/l.

**Approximate reference values**
23—31 mmol/l.

**Notes**
1   May be affected by respiratory,
non-respiratory and compensatory
factors.
2   It is an insensitive indicator of
abnormalities of acid-base balance.

**Conversion**

Nil — as bicarbonate.

Standard bicarbonate is derived from measurements in whole blood equilibrated at a $P\ CO_2$ of 5.3 kPa (40 mmHg) at $37°C$. It is intended to provide a measure of bicarbonate concentration unaffected by purely respiratory factors.

**Approximate reference values**

22—26 mmol/l.

**Notes**

1    This test requires arterial or free flowing capillary blood (see Hydrogen ion).

2    Standard bicarbonate measures the non-respiratory component in a disorder of acid-base balance. We find it a more reliable test than serum or plasma bicarbonate. However it **does not** distinguish between primary non-respiratory disorders and secondary compensation for a primary respiratory disorder and should always be interpreted with $H^+$ and $P\ CO_2$ and in the light of the clinical picture.

# Serum bilirubin (total) (μmol/l)

**Conversion**

$$\frac{\mu mol/l}{17.1} = mg/100\ ml$$

**Approximate reference values**

*'Healthy', ambulant*

| Male | Female |
|------|--------|
| 3–21 | 3–21 |

*Hospital in-patients*

| Male | Female |
|------|--------|
| 3–26 | 3–21 |

No significant differences with age.

**Note**
Bilirubin fades rapidly in bright light leading to a falsely low serum bilirubin.

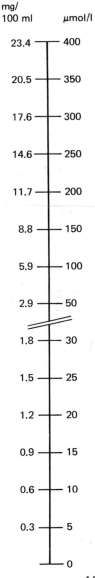

| mg/100 ml | μmol/l |
|-----------|--------|
| 23.4 | 400 |
| 20.5 | 350 |
| 17.6 | 300 |
| 14.6 | 250 |
| 11.7 | 200 |
| 8.8 | 150 |
| 5.9 | 100 |
| 2.9 | 50 |
| 1.8 | 30 |
| 1.5 | 25 |
| 1.2 | 20 |
| 0.9 | 15 |
| 0.6 | 10 |
| 0.3 | 5 |
| | 0 |

# Serum calcium (mmol/l)

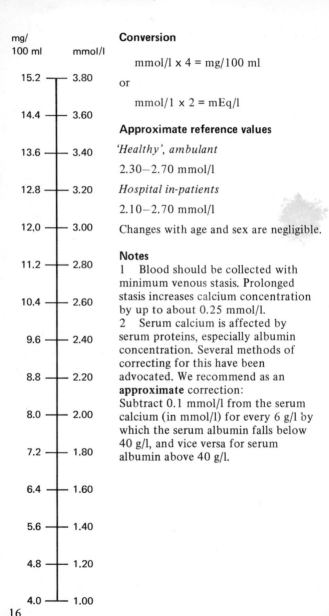

| mg/100 ml | mmol/l |
|-----------|--------|
| 15.2 | 3.80 |
| 14.4 | 3.60 |
| 13.6 | 3.40 |
| 12.8 | 3.20 |
| 12.0 | 3.00 |
| 11.2 | 2.80 |
| 10.4 | 2.60 |
| 9.6 | 2.40 |
| 8.8 | 2.20 |
| 8.0 | 2.00 |
| 7.2 | 1.80 |
| 6.4 | 1.60 |
| 5.6 | 1.40 |
| 4.8 | 1.20 |
| 4.0 | 1.00 |

**Conversion**

mmol/l × 4 = mg/100 ml

or

mmol/l × 2 = mEq/l

**Approximate reference values**

*'Healthy', ambulant*

2.30–2.70 mmol/l

*Hospital in-patients*

2.10–2.70 mmol/l

Changes with age and sex are negligible.

**Notes**
1   Blood should be collected with minimum venous stasis. Prolonged stasis increases calcium concentration by up to about 0.25 mmol/l.
2   Serum calcium is affected by serum proteins, especially albumin concentration. Several methods of correcting for this have been advocated. We recommend as an **approximate** correction:
Subtract 0.1 mmol/l from the serum calcium (in mmol/l) for every 6 g/l by which the serum albumin falls below 40 g/l, and vice versa for serum albumin above 40 g/l.

16

# Serum chloride (mmol/l)

**Conversion**
Nil. For chloride, mmol/l = mEq/l.

**Approximate reference values**
98–108 mmol/l.

**Note**
1    Serum or plasma chloride is
affected by a wide variety of diseases,
including disorders of water balance,
sodium balance, acid-base balance, and
renal function. In our view it is
obsolete as part of routine 'electrolytes'.

# Serum cholesterol (mmol/l)

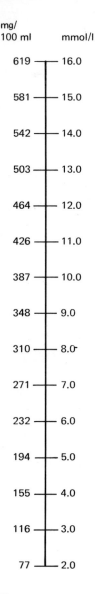

| mg/100 ml | mmol/l |
|---|---|
| 619 | 16.0 |
| 581 | 15.0 |
| 542 | 14.0 |
| 503 | 13.0 |
| 464 | 12.0 |
| 426 | 11.0 |
| 387 | 10.0 |
| 348 | 9.0 |
| 310 | 8.0 |
| 271 | 7.0 |
| 232 | 6.0 |
| 194 | 5.0 |
| 155 | 4.0 |
| 116 | 3.0 |
| 77 | 2.0 |

## Conversion

mmol/l x 38.7 = mg/100 ml

## Approximate reference values

### 'Healthy', ambulant

| Age | Male | Female |
|---|---|---|
| 20–29 | 3.5–7.2 | 3.4–7.1 |
| 30–39 | 3.9–8.3 | 3.6–7.8 |
| 40–49 | 4.2–8.5 | 4.1–8.1 |
| 50–59 | 4.2–8.5 | 4.4–9.2 |
| 60–69 | 4.3–8.5 | 5.0–9.1 |
| 70+ | 2.9–8.7 | 3.9–9.1 |

### Hospital in-patients

| 20–29 | 3.0–7.7 | 3.3–7.9 |
|---|---|---|
| 30–39 | 3.1–8.2 | 3.4–7.9 |
| 40–49 | 3.5–8.7 | 3.5–8.7 |
| 50–59 | 3.3–8.5 | 3.4–9.3 |
| 60–69 | 3.0–8.6 | 3.4–9.5 |
| 70+ | 3.1–8.1 | 3.4–9.3 |

## Note

Appropriate reference values for serum cholesterol are controversial, since some authorities believe that an 'undesirably high' serum cholesterol is common in the absence of demonstrable disease. We believe that every value above 8.0 mmol/l should be treated as abnormal.

# Serum cortisol (nmol/l)

**Conversion**

$$\frac{\text{nmol/l}}{27.6} = \mu g/100 \text{ ml}$$

**Approximate reference values**

*At 0900 hrs*

140–700 nmol/l

*At 2400 hrs*

Less than 140 nmol/l

**Notes**

1   There is a marked diurnal variation. Careful timing (and labelling!) of specimens is essential.

2   Non-specific fluorimetric methods are still widely used. These also measure other 11 hydroxy corticosteroids, and are prone to error if blood is not separated promptly.

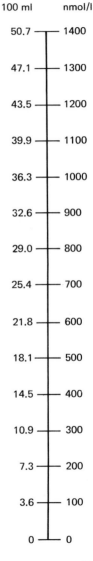

| μg/100 ml | nmol/l |
|---|---|
| 50.7 | 1400 |
| 47.1 | 1300 |
| 43.5 | 1200 |
| 39.9 | 1100 |
| 36.3 | 1000 |
| 32.6 | 900 |
| 29.0 | 800 |
| 25.4 | 700 |
| 21.8 | 600 |
| 18.1 | 500 |
| 14.5 | 400 |
| 10.9 | 300 |
| 7.3 | 200 |
| 3.6 | 100 |
| 0 | 0 |

# Serum creatinine (μmol/l)

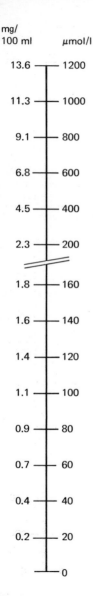

| mg/<br>100 ml | μmol/l |
|---|---|
| 13.6 | 1200 |
| 11.3 | 1000 |
| 9.1 | 800 |
| 6.8 | 600 |
| 4.5 | 400 |
| 2.3 | 200 |
| 1.8 | 160 |
| 1.6 | 140 |
| 1.4 | 120 |
| 1.1 | 100 |
| 0.9 | 80 |
| 0.7 | 60 |
| 0.4 | 40 |
| 0.2 | 20 |
|  | 0 |

## Conversion

$$\frac{\mu mol/l}{88.4} = mg/100\ ml$$

## Approximate reference values

*'Healthy', ambulant*

| Age | Male | Female |
|---|---|---|
| 20–29 | 60–120 | 40–100 |
| 30–39 | 60–120 | 50–100 |
| 40–49 | 60–120 | 45–100 |
| 50–59 | 60–125 | 50–100 |
| 60–69 | 60–125 | 50–110 |
| 70+ | 60–135 | 45–110 |

*Hospital in-patients*

| Age | Male | Female |
|---|---|---|
| 20–29 | 65–130 | 60–115 |
| 30–39 | 70–140 | 60–115 |
| 40–49 | 70–145 | 55–110 |
| 50–59 | 65–140 | 60–120 |
| 60–69 | 70–155 | 60–125 |
| 70+ | 70–165 | 60–130 |

## Note

Serum creatinine may remain within normal limits down to a glomerular filtration rate of about 60 ml/min.

# Blood gases, carbon dioxide ($P\,CO_2$) and oxygen ($P\,O_2$), kPa

**Conversion**

kPa x 7.5 = mmHg.

**Approximate reference values**

$P\,CO_2$

4.7–6.0 kPa

$P\,O_2$

11.3–14.0 kPa

**Notes**

1    Ideally, arterial blood taken anaerobically into a heparinised syringe should be used. *Free flowing* capillary blood, collected carefully in special tubes may be used for **$P\,CO_2$**, but is **not satisfactory** for **$P\,O_2$**.

2    Blood **must** be taken to the laboratory promptly, especially for $P\,O_2$.

3    Interpretation of $P\,O_2$ requires allowance for the oxygen content of the air the patient was breathing when the blood was collected.

| mmHg | kPa |
|---|---|
| 150 | 20.0 |
| 135 | 18.0 |
| 120 | 16.0 |
| 105 | 14.0 |
| 90 | 12.0 |
| 75 | 10.0 |
| 60 | 8.0 |
| 45 | 6.0 |
| 30 | 4.0 |
| 15 | 2.0 |
| 0 | 0 |

# Serum globulin (g/l)

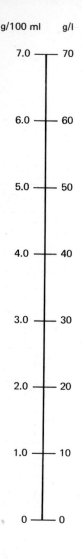

| g/100 ml | g/l |
|----------|-----|
| 7.0 | 70 |
| 6.0 | 60 |
| 5.0 | 50 |
| 4.0 | 40 |
| 3.0 | 30 |
| 2.0 | 20 |
| 1.0 | 10 |
| 0 | 0 |

**Conversion**

$$\frac{g/l}{10} = g/100 \text{ ml}$$

**Approximate reference values**

*'Healthy', ambulant*

21–37 g/l

*Hospital in-patients*

20–42 g/l

No significant changes with age or sex.

**Note**
Serum globulin is usually measured as the difference between total protein and albumin. These values are based upon values for albumin determined by the automatic brom-cresol green method.
   **Check with your local laboratory.**

## Conversion

mmol/l x 18 = mg/100 ml

## Approximate reference values

*'Healthy', ambulant*

| Age | Male | Female |
|-----|------|--------|
| 20–29 | 3.4–6.7 | 3.5–6.7 |
| 30–39 | 3.5–6.7 | 3.5–6.7 |
| 40–49 | 3.4–7.0 | 3.4–7.0 |
| 50–59 | 3.6–7.1 | 3.6–7.1 |
| 60–69 | 3.3–7.4 | 3.4–7.4 |
| 70+ | 2.9–7.5 | 2.9–7.5 |

*Hospital in-patients*

| Age | Male | Female |
|-----|------|--------|
| 20–29 | 3.7–8.3 | 3.4–7.8 |
| 30–39 | 3.6–9.1 | 3.5–8.3 |
| 40–49 | 3.4–9.7 | 3.6–8.9 |
| 50–59 | 3.7–10.3 | 3.8–9.8 |
| 60–69 | 4.1–11.9 | 4.1–11.9 |
| 70+ | 3.9–11.4 | 4.3–12.9 |

## Notes

1 These values are for random (i.e. non-fasting) glucose. Doubtful values may need following up with a fasting blood glucose or glucose tolerance test.
2 For accurate blood glucose, blood should normally be collected in a tube containing a preservative such as sodium fluoride.
3 **Fasting** glucose values are usually in the range 3.0–5.3 mmol/l, approximately.
4 Serum glucose is approximately 12% higher than whole blood glucose.
5 Laboratories using non-specific methods measuring other 'saccharoids' may report values about 1.0 mmol/l higher than those quoted here.

| mg/100 ml | mmol/l |
|-----------|--------|
| 1260 | 70.0 |
| 1080 | 60.0 |
| 900 | 50.0 |
| 720 | 40.0 |
| 540 | 30.0 |
| 360 | 20.0 |
| 270 | 15.0 |
| 252 | 14.0 |
| 234 | 13.0 |
| 216 | 12.0 |
| 198 | 11.0 |
| 180 | 10.0 |
| 162 | 9.0 |
| 144 | 8.0 |
| 126 | 7.0 |
| 108 | 6.0 |
| 90 | 5.0 |
| 72 | 4.0 |
| 54 | 3.0 |
| 36 | 2.0 |
| 18 | 1.0 |
|  | 0 |

# Blood hydrogen ion (H⁺) concentration (nmol/l)

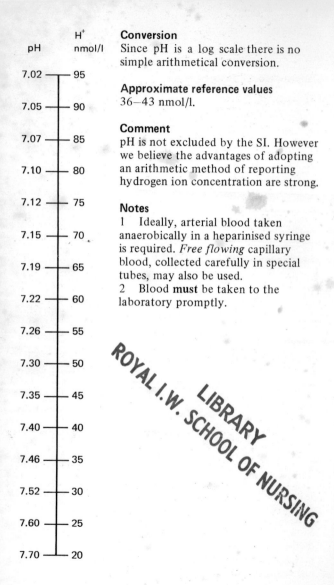

| pH | H⁺ nmol/l |
|------|------|
| 7.02 | 95 |
| 7.05 | 90 |
| 7.07 | 85 |
| 7.10 | 80 |
| 7.12 | 75 |
| 7.15 | 70 |
| 7.19 | 65 |
| 7.22 | 60 |
| 7.26 | 55 |
| 7.30 | 50 |
| 7.35 | 45 |
| 7.40 | 40 |
| 7.46 | 35 |
| 7.52 | 30 |
| 7.60 | 25 |
| 7.70 | 20 |

**Conversion**
Since pH is a log scale there is no simple arithmetical conversion.

**Approximate reference values**
36–43 nmol/l.

**Comment**
pH is not excluded by the SI. However we believe the advantages of adopting an arithmetic method of reporting hydrogen ion concentration are strong.

**Notes**
1   Ideally, arterial blood taken anaerobically in a heparinised syringe is required. *Free flowing* capillary blood, collected carefully in special tubes, may also be used.
2   Blood **must** be taken to the laboratory promptly.

# Serum iron and iron-binding capacity ($\mu$mol/l)

**Conversion**

$\mu$mol/l x 5.6 = $\mu$g/100 ml.

**Approximate reference values Iron**

*'Healthy', ambulant*

| Age | Male | Female |
|-----|------|--------|
| 20–29 | 8–31 | 6–29 |
| 30–39 | 8–28 | 3–30 |
| 40–49 | 8–28 | 5–29 |
| 50–59 | 7–29 | 7–26 |
| 60–69 | 6–29 | 7–26 |
| 70+ | 6–29 | 6–26 |

*Hospital in-patients*

| Age | Male | Female |
|-----|------|--------|
| 20–29 | 3–29 | 2–34 |
| 30–39 | 2–38 | 2–32 |
| 40–49 | 2–36 | 2–32 |
| 50–59 | 2–32 | 2–29 |
| 60–69 | 2–30 | 2–28 |
| 70+ | 2–30 | 2–26 |

**Iron-binding capacity**
45–72 $\mu$mol/l

**Notes**
Serum iron is commonly misinterpreted.
1   It is higher in the morning than evening, and preferably blood should be collected between 0900 and 1000 hours.
2   It is reduced in a variety of diseases as part of a response to inflammation without any concomitant deficiency of iron.

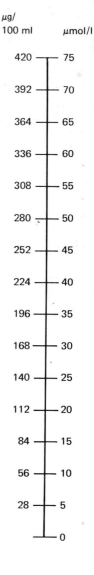

| $\mu$g/100 ml | $\mu$mol/l |
|-----|-----|
| 420 | 75 |
| 392 | 70 |
| 364 | 65 |
| 336 | 60 |
| 308 | 55 |
| 280 | 50 |
| 252 | 45 |
| 224 | 40 |
| 196 | 35 |
| 168 | 30 |
| 140 | 25 |
| 112 | 20 |
| 84 | 15 |
| 56 | 10 |
| 28 | 5 |
| | 0 |

# Serum magnesium (mmol/l)

| mEq/l | mmol/l |
|-------|--------|
| 4.0 | 2.0 |
| 3.6 | 1.8 |
| 3.2 | 1.6 |
| 2.8 | 1.4 |
| 2.4 | 1.2 |
| 2.0 | 1.0 |
| 1.6 | 0.8 |
| 1.2 | 0.6 |
| 0.8 | 0.4 |
| 0.4 | 0.2 |
| 0 | 0 |

**Conversion**

mmol/l x 2
= mEq/l

or

mmol/l x 2.4
= mg/100 ml.

**Approximate reference values**
0.70–0.95 mmol/l

**Note**
These values are for determination of serum magnesium by atomic absorption spectroscopy. Less specific methods tend to give higher values.
**Check with your local laboratory**.

| mmol/l | mg/100 ml |
|--------|-----------|
| 2.0 | 4.8 |
| 1.8 | 4.3 |
| 1.6 | 3.8 |
| 1.4 | 3.4 |
| 1.2 | 2.9 |
| 1.0 | 2.4 |
| 0.8 | 1.9 |
| 0.6 | 1.4 |
| 0.4 | 1.0 |
| 0.2 | 0.5 |
| 0 | |

# Serum phosphate, inorganic (mmol/l)

**Conversion**

   mmol/l x 3.1 = mg/100 ml.

**Approximate reference values**

| Male | Female |
|------|--------|
| 0.7–1.4 mmol/l | 0.8–1.4 mmol/l |

**Notes**

1   Serum inorganic phosphate tends to fall after a meal and blood should be collected after an overnight fast.

2   Haemolysed serum, or serum left in prolonged contact with red cells, has falsely high phosphate concentration.

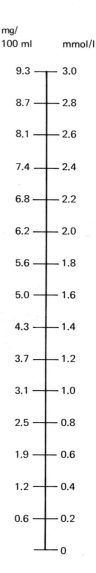

| mg/100 ml | mmol/l |
|-----------|--------|
| 9.3 | 3.0 |
| 8.7 | 2.8 |
| 8.1 | 2.6 |
| 7.4 | 2.4 |
| 6.8 | 2.2 |
| 6.2 | 2.0 |
| 5.6 | 1.8 |
| 5.0 | 1.6 |
| 4.3 | 1.4 |
| 3.7 | 1.2 |
| 3.1 | 1.0 |
| 2.5 | 0.8 |
| 1.9 | 0.6 |
| 1.2 | 0.4 |
| 0.6 | 0.2 |
|  | 0 |

# Serum protein-bound iodine (P.B.I.) nmol/l

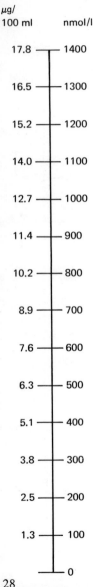

| µg/100 ml | nmol/l |
|---|---|
| 17.8 | 1400 |
| 16.5 | 1300 |
| 15.2 | 1200 |
| 14.0 | 1100 |
| 12.7 | 1000 |
| 11.4 | 900 |
| 10.2 | 800 |
| 8.9 | 700 |
| 7.6 | 600 |
| 6.3 | 500 |
| 5.1 | 400 |
| 3.8 | 300 |
| 2.5 | 200 |
| 1.3 | 100 |
|  | 0 |

**Conversion**

$$\frac{\text{nmol/l}}{78.8} = \text{µg/100 ml}$$

**Approximate reference values**
280–630 nmol/l

**Notes**
1   Values, even in nmol/l, are **not** interchangeable with thyroxine.
2   P.B.I. can be grossly increased by iodides in cough mixtures etc., drugs such as enterovioform and iodine containing X-ray contrast media, e.g. after I.V.P., arteriogram, etc.
3   Moderate increases are caused by pregnancy, or oestrogens including the 'pill' due to increases in thyroxine-binding globulin.

# Serum sodium and potassium (mmol/l)

**Conversion**
Nil. For sodium and potassium
mmol/l = mEq/1.

**Approximate reference values**

Sodium

*'Healthy', ambulant*

135—146 mmol/l

*Hospital in-patients*

131—146 mmol/l

No significant changes with age or sex.

Potassium

*'Healthy', ambulant*

| Male | Female |
|------|--------|
| 3.7—5.2 mmol/l | 3.5—5.1 mmol/l |

*Hospital in-patients*

| Male | Female |
|------|--------|
| 3.4—5.4 mmol/l | 3.3—5.3 mmol/l |

No significant changes with age.

**Notes**
1   Abnormalities of serum sodium
*concentration* are not always due to
abnormal sodium balance, but are often
due to excess or deficiency of water.
2   Blood for potassium must be
separated promptly. Long delay,
especially if blood is refrigerated, or
haemolysis cause large increases in
serum potassium.
3   Plasma potassium values tend to be
a little lower than values on serum.

# Serum thyroxine ('T₄') nmol/l

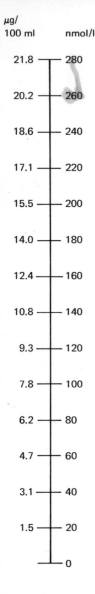

| μg/ 100 ml | nmol/l |
|---|---|
| 21.8 | 280 |
| 20.2 | 260 |
| 18.6 | 240 |
| 17.1 | 220 |
| 15.5 | 200 |
| 14.0 | 180 |
| 12.4 | 160 |
| 10.8 | 140 |
| 9.3 | 120 |
| 7.8 | 100 |
| 6.2 | 80 |
| 4.7 | 60 |
| 3.1 | 40 |
| 1.5 | 20 |
|  | 0 |

## Conversion

$$\frac{\text{nmol/l}}{12.87} = \mu g/100 \text{ ml}$$

## Approximate reference values
60–135 nmol/l.

## Notes
1 Tends to be moderately increased by pregnancy or oestrogens including the 'pill' due to increased thyroxine-binding globulin.
2 Tends to be decreased by hypoproteinaemia due to malnutrition, malabsorption etc., or drugs such as phenylbutazone.
3 **Caution!** Some laboratories report thyroxine iodine – values are similar to protein bound iodine but quite different from thyroxine.

**Conversion**
Nil. Units are arbitrary.

**Approximate reference values**
93—117.

**Notes**
This test is confusing.
1   It does not measure triiodothyronine
($T_3$) and should not be abbreviated to
'$T_3$' since triiodothyronine can be
measured specifically.
2   It is reported in two opposite
conventions. We prefer the convention
by which high values indicate many
free thyroxine binding protein binding
sites (as in hypothyroidism or with high
serum thyroxine binding globulin as
occurs in pregnancy or patients on the
pill). Low values indicate reduced free
binding sites (as in hyperthyroidism or
hypoproteinaemia). The $T_3$ resin uptake
convention is the reverse of this.
    **Check with your local laboratory.**

# Serum triglycerides (mmol/l) fasting

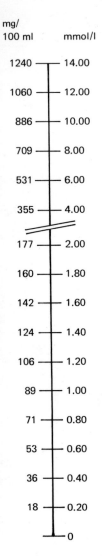

| mg/ 100 ml | mmol/l |
|---|---|
| 1240 | 14.00 |
| 1060 | 12.00 |
| 886 | 10.00 |
| 709 | 8.00 |
| 531 | 6.00 |
| 355 | 4.00 |
| 177 | 2.00 |
| 160 | 1.80 |
| 142 | 1.60 |
| 124 | 1.40 |
| 106 | 1.20 |
| 89 | 1.00 |
| 71 | 0.80 |
| 53 | 0.60 |
| 36 | 0.40 |
| 18 | 0.20 |
|  | 0 |

**Conversion**
mmol/l x 88.6 = mg/100 ml

**Approximate reference values**
0.80–1.80 mmol/l .

**Note**
Blood **must** be taken after a fast of at least twelve hours.

# Serum urate (μmol/l)
## (formerly termed uric acid)

**Conversion**

$$\frac{\mu mol/l}{59.5} = mg/100\ ml$$

**Approximate reference values**

*'Healthy', ambulant*

| Age | Male | Female |
|-----|------|--------|
| 20–29 | 240–510 | 170–410 |
| 30–39 | 250–530 | 150–400 |
| 40–49 | 245–530 | 155–400 |
| 50–59 | 245–530 | 170–440 |
| 60–69 | 240–530 | 195–470 |
| 70+ | 180–580 | 160–460 |

*Hospital in-patients*

| Age | Male | Female |
|-----|------|--------|
| 20–29 | 215–555 | 150–410 |
| 30–39 | 200–575 | 160–440 |
| 40–49 | 200–525 | 155–455 |
| 50–59 | 200–545 | 165–430 |
| 60–69 | 185–565 | 160–480 |
| 70+ | 195–570 | 160–540 |

**Notes**

1   Impaired renal function may raise serum urate in the absence of a primary disorder of urate metabolism.
2   Some laboratories report serum urate in mmol/l – i.e. the values are one thousandth of those expressed in μmol/l.

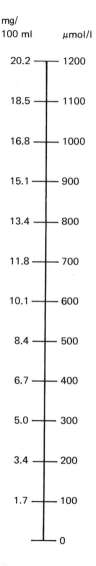

| mg/100 ml | μmol/l |
|-----------|--------|
| 20.2 | 1200 |
| 18.5 | 1100 |
| 16.8 | 1000 |
| 15.1 | 900 |
| 13.4 | 800 |
| 11.8 | 700 |
| 10.1 | 600 |
| 8.4 | 500 |
| 6.7 | 400 |
| 5.0 | 300 |
| 3.4 | 200 |
| 1.7 | 100 |
| | 0 |

# Serum urea (mmol/l)

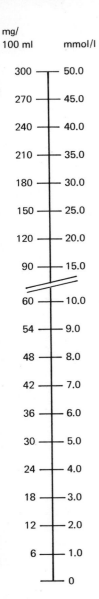

| mg/ 100 ml | mmol/l |
|---|---|
| 300 | 50.0 |
| 270 | 45.0 |
| 240 | 40.0 |
| 210 | 35.0 |
| 180 | 30.0 |
| 150 | 25.0 |
| 120 | 20.0 |
| 90 | 15.0 |
| 60 | 10.0 |
| 54 | 9.0 |
| 48 | 8.0 |
| 42 | 7.0 |
| 36 | 6.0 |
| 30 | 5.0 |
| 24 | 4.0 |
| 18 | 3.0 |
| 12 | 2.0 |
| 6 | 1.0 |
|  | 0 |

## Conversion

mmol/l x 6 = mg/100 ml

## Approximate reference values

*'Healthy', ambulant*

| Age | Male | Female |
|---|---|---|
| 20–29 | 3.3–7.5 | 2.7–7.3 |
| 30–39 | 3.7–7.5 | 3.2–6.7 |
| 40–49 | 3.7–7.5 | 3.2–7.2 |
| 50–59 | 3.7–7.7 | 3.5–7.7 |
| 60–69 | 3.8–8.2 | 3.7–8.0 |
| 70+ | 3.5–8.5 | 3.6–7.3 |

*Hospital in-patients*

| Age | Male | Female |
|---|---|---|
| 20–29 | 2.7–8.8 | 2.2–6.7 |
| 30–39 | 2.8–8.3 | 2.4–7.3 |
| 40–49 | 2.7–9.2 | 2.5–7.7 |
| 50–59 | 3.0–9.5 | 2.8–8.7 |
| 60–69 | 3.1–10.4 | 2.8–9.5 |
| 70+ | 3.5–12.3 | 3.0–10.7 |

In practice, we feel any patient with a serum urea above 8.0 mmol/l should be further investigated.

## Notes

1   Like serum creatinine, serum urea may remain normal down to a glomerular filtration rate of about 60 ml/min.
2   Unlike serum creatinine, it is significantly affected by dietary protein intake.

# C.S.F. glucose (mmol/l)

**Conversion**
mmol/l x 18 = mg/100 ml

**Approximate reference values**
3.3–4.4 mmol/l.

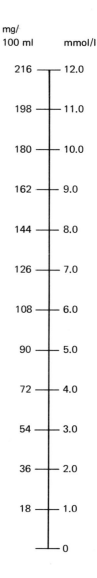

| mg/100 ml | mmol/l |
|---|---|
| 216 | 12.0 |
| 198 | 11.0 |
| 180 | 10.0 |
| 162 | 9.0 |
| 144 | 8.0 |
| 126 | 7.0 |
| 108 | 6.0 |
| 90 | 5.0 |
| 72 | 4.0 |
| 54 | 3.0 |
| 36 | 2.0 |
| 18 | 1.0 |
|  | 0 |

## C.S.F. protein (g/l)

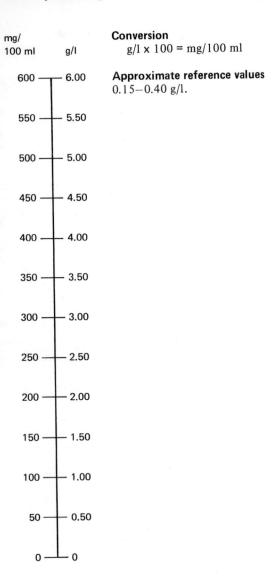

| mg/100 ml | g/l |
|---|---|
| | |

**Conversion**
g/l x 100 = mg/100 ml

**Approximate reference values**
0.15–0.40 g/l.

| mg/100 ml | g/l |
|---|---|
| 600 | 6.00 |
| 550 | 5.50 |
| 500 | 5.00 |
| 450 | 4.50 |
| 400 | 4.00 |
| 350 | 3.50 |
| 300 | 3.00 |
| 250 | 2.50 |
| 200 | 2.00 |
| 150 | 1.50 |
| 100 | 1.00 |
| 50 | 0.50 |
| 0 | 0 |

# Urine calcium and phosphate (mmol/24h)

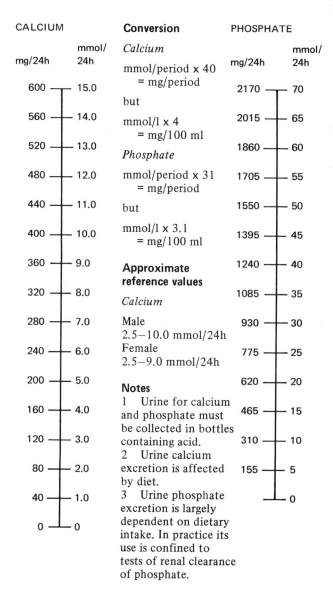

| CALCIUM | | Conversion | PHOSPHATE | |
|---|---|---|---|---|
| mg/24h | mmol/24h | | mg/24h | mmol/24h |

**Conversion**

*Calcium*

mmol/period x 40 = mg/period

but

mmol/l x 4 = mg/100 ml

*Phosphate*

mmol/period x 31 = mg/period

but

mmol/l x 3.1 = mg/100 ml

**Approximate reference values**

*Calcium*

Male
2.5–10.0 mmol/24h
Female
2.5–9.0 mmol/24h

**Notes**

1 Urine for calcium and phosphate must be collected in bottles containing acid.

2 Urine calcium excretion is affected by diet.

3 Urine phosphate excretion is largely dependent on dietary intake. In practice its use is confined to tests of renal clearance of phosphate.

CALCIUM scale:
600 — 15.0
560 — 14.0
520 — 13.0
480 — 12.0
440 — 11.0
400 — 10.0
360 — 9.0
320 — 8.0
280 — 7.0
240 — 6.0
200 — 5.0
160 — 4.0
120 — 3.0
80 — 2.0
40 — 1.0
0 — 0

PHOSPHATE scale:
2170 — 70
2015 — 65
1860 — 60
1705 — 55
1550 — 50
1395 — 45
1240 — 40
1085 — 35
930 — 30
775 — 25
620 — 20
465 — 15
310 — 10
155 — 5
0 — 0

## Urine creatinine (mmol/period) and creatinine clearance

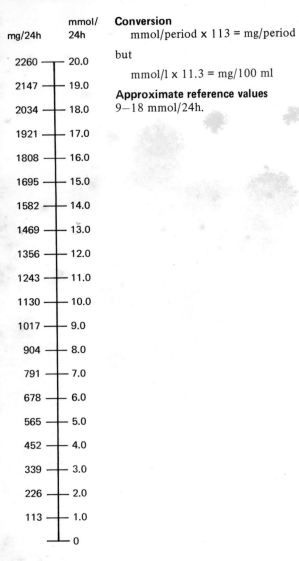

| mg/24h | mmol/24h |
|---|---|
| 2260 | 20.0 |
| 2147 | 19.0 |
| 2034 | 18.0 |
| 1921 | 17.0 |
| 1808 | 16.0 |
| 1695 | 15.0 |
| 1582 | 14.0 |
| 1469 | 13.0 |
| 1356 | 12.0 |
| 1243 | 11.0 |
| 1130 | 10.0 |
| 1017 | 9.0 |
| 904 | 8.0 |
| 791 | 7.0 |
| 678 | 6.0 |
| 565 | 5.0 |
| 452 | 4.0 |
| 339 | 3.0 |
| 226 | 2.0 |
| 113 | 1.0 |
|  | 0 |

**Conversion**

mmol/period x 113 = mg/period

but

mmol/l x 11.3 = mg/100 ml

**Approximate reference values**
9–18 mmol/24h.

This is derived from the formula

$$\frac{U \times V \times 1000}{S \times T}$$

and is expressed as ml/min, where

U = urine creatinine concentration in mmol/l;
V = urine volume in ml;
S = serum creatinine concentration in $\mu$mol/l
T = time of urine collection in minutes.

## Approximate reference values

Male              Female
85–125 ml/min     75–115 ml/min.

## Notes

1    Creatinine clearance values are approximately 10% higher if a specific method for determining creatinine is used.

2    The major source of error is inaccurate urine collection. We recommend 2 x 24 hour urine collections with at least one serum specimen at some time during the test.

3    Since creatinine clearance depends amongst other factors on body size, for unusually large or small patients the body surface area should be determined from nomograms for height and weight, and creatinine clearance corrected by

$$\times \frac{1.73}{\text{surface area in m}^2}$$

## Urine H.M.M.A. (μmol/period)
## (4-hydroxy-3-methoxy mandelic acid or 'V.M.A.')

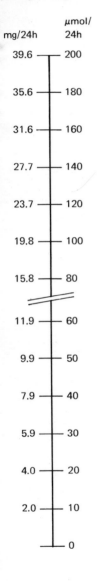

| mg/24h | μmol/24h |
|--------|----------|
| 39.6 | 200 |
| 35.6 | 180 |
| 31.6 | 160 |
| 27.7 | 140 |
| 23.7 | 120 |
| 19.8 | 100 |
| 15.8 | 80 |
| 11.9 | 60 |
| 9.9 | 50 |
| 7.9 | 40 |
| 5.9 | 30 |
| 4.0 | 20 |
| 2.0 | 10 |
| | 0 |

**Conversion**

$$\frac{\mu mol/period}{5.05} = mg/period.$$

**Approximate reference values**
10—35 μmol/24h.

**Notes**
1    Urine **must** be collected in bottles containing acid.
2    Ideally patients should be taken off all medication at least 48 hrs before the urine collection is started. Many drugs may interfere with this investigation.
3    The importance of restricting dietary sources of vanillin has been over emphasised.

# Urine metanephrines (μmol/period)

**Conversion (as normetadrenaline)**

$$\frac{\mu\text{mol/period}}{5.46} = \text{mg/period}$$

**Approximate reference values**
0.5–7.0 μmol/24h.

**Notes**
1  Urine **must** be collected in bottles containing acid.
2  Ideally patients should be taken off all medication at least two days before the urine collection is started.

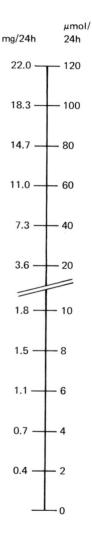

| mg/24h | μmol/24h |
|---|---|
| 22.0 | 120 |
| 18.3 | 100 |
| 14.7 | 80 |
| 11.0 | 60 |
| 7.3 | 40 |
| 3.6 | 20 |
| 1.8 | 10 |
| 1.5 | 8 |
| 1.1 | 6 |
| 0.7 | 4 |
| 0.4 | 2 |
| | 0 |

# Urine area (mmol/period)

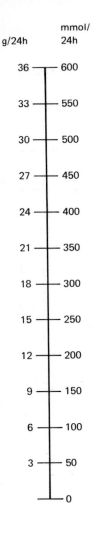

| g/24h | mmol/24h |
|-------|----------|
| 36 | 600 |
| 33 | 550 |
| 30 | 500 |
| 27 | 450 |
| 24 | 400 |
| 21 | 350 |
| 18 | 300 |
| 15 | 250 |
| 12 | 200 |
| 9 | 150 |
| 6 | 100 |
| 3 | 50 |
| | 0 |

## Conversion

$$\frac{mmol/period}{16.6} = g/period$$

but

$$\frac{mmol/l}{166} = g/100\ ml$$

## Approximate reference values

250–600 mmol/24h.

## Note

Urea excretion is dependent on 3 main factors — renal function, protein intake and 'catabolism'.

# Faecal (stool) fat (mmol/period)

**Conversion**

$$\frac{\text{mmol}/24\text{h}}{3.52} = \text{g}/24\text{h}$$

**Approximate reference values**
11–18 mmol/24h.

**Notes**
1  The calculation assumes an average molecular weight of 284 for faecal fatty acids. This is not appropriate if the patient is treated with medium chain triglycerides.
2  Faecal collections are so erratic that faeces should be collected over at least three days.
3  Ensure the patient's fat intake is at least 50g per day.

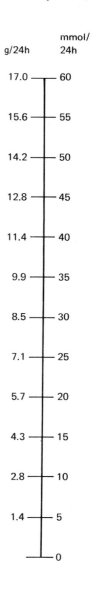

| g/24h | mmol/24h |
|---|---|
| 17.0 | 60 |
| 15.6 | 55 |
| 14.2 | 50 |
| 12.8 | 45 |
| 11.4 | 40 |
| 9.9 | 35 |
| 8.5 | 30 |
| 7.1 | 25 |
| 5.7 | 20 |
| 4.3 | 15 |
| 2.8 | 10 |
| 1.4 | 5 |
|  | 0 |

# Serum enzymes

The use of International Units in clinical enzymology does not completely clarify the expression of enzyme activity in serum. All enzyme assay methods are dependent on the substrate, technique and temperature employed. The following Reference Ranges are quoted in conventional and/or International Units (IU/l) for enzymes in common use when assayed at 37°. For some assays the use of International Units is common practice. **Great care should be taken to establish which method has been used and to ensure that the ranges quoted are compatible with those used by your laboratory.**

| Enzymes | Method of determination | Reference values determined at 37° | | References (p. 47) |
| | | Conventional units | International units | |
|---|---|---|---|---|
| 1. Acid Phosphatase | | | | |
| a) Total | King–Armstrong | Less than 4.6 | Less than 8.2 IU/l | 1a |
| b) Prostatic (Formaldehyde stable) | King–Armstrong | Less than 4.0 | Less than 7.2 IU/l | 1b |
| 2. Alkaline Phosphatase | King–Armstrong | 3–14 | 21–100 IU/l | 1a |
| | Bessey–Lowry | 0–3 | 20–50 IU/l | 2a |
| | Bodansky | 1.5–5 | 8–27 IU/l | 2b |

| Test | Method | | IU/l | Ref |
|---|---|---|---|---|
| 3. Amylase | Somogyi | 80–180 | 150–340 IU/l | 3a, 3b |
| | Street–Close | 9–35 | | 3c |
| | Dyed substrates – | | | |
| | Phadebas | 70–300 | | 3d |
| | Amylochrome | 45–200 | | 3e |
| | Dy-Amyl | 36–158 | | 3f |
| 4. Alanine transaminase (Ala-T or SGPT) | Reitman–Frankel | | 5–30 IU/l | 4a, 4b |
| 5. Aspartate transaminase (Asp-T or SGOT) | Reitman–Frankel or Karmen | 5–40 | 5–30 IU/l | 4a, 5a, 5b |
| 6. Creatine kinase (CK or CPK) | Oliver (modified) Rosalki | | Less than 130 IU/l ♂ Less than 100 IU/l ♀ Less than 65 IU/l | 6a 6b |
| 7. 2-Hydroxybutyrate dehydrogenase (HBD) | Rosalki–Wilkinson McQueen | 320–675 | 150–325 IU/l 80–440 IU/l | 7a 7b |

**Serum enzymes — Continued**

| Enzymes | Method of determination | Reference values determined at 37° | | References (p. 47) |
| | | Conventional units | International units | |
|---|---|---|---|---|
| 8. Lactate dehydrogenase (LDH) | Wroblewski–La Due McQueen | 250–760 | 120–365 IU/l 240–525 IU/l | 8a, 8b 7b |
| 9. γ-Glutamyl transpeptidase (γ-GT) | Szasz | | ♂ 6–28 IU/l ♀ 4–18 IU/l | 9 |
| 10. 5-Nucleotidase (5-NT) | Persijn | | 4–14 IU/l | 10 |